FLORA OF TROPICAL EAST AFRICA

CANELLACEAE

B. Verdcourt

(East African Herbarium)

Aromatic glabrous trees. Leaves simple, gland-dotted, alternate, without stipules. Inflorescences axillary or axillary and terminal, cymose ; bracts * three, persistent. Flowers regular, hermaphrodite. Sepals 4–5, free, rather thick, imbricate. Petals 4–12, free or connate, thin, imbricate or wanting. Stamens hypogynous, up to 20 ; filaments united into a tube ; anthers extrorse, adnate to the upper part of the tube. Ovary superior, 1–locular ; ovules 2–many, subanatropous; placentation parietal, placentae 2–5. Style thick and short ; stigmas 2–5. Fruit a berry. Seeds 2–many, shining, with oily and fleshy endosperm.

The family is represented by only one genus on the African mainland, of which all the known species occur in East Africa.

WARBURGIA

Engl., P.O.A. C : 276 (1895)

Evergreen trees. Leaves very shining above. Inflorescences axillary, flowers solitary or 3–4 in reduced cymes ; bracts ovate or reniform, very obtuse. Sepals 5, ovate, obtuse. Petals 5, obovate-spathulate, obtuse, narrowed to the base. Stamens 10, united into a tube, the part extending above the anthers 10-crenellated, anthers 10. Ovary narrowly ellipsoidal, narrowed towards the apex, clavate at the base ; ovules 10–31, biseriate on 5 placentae. Style inverted cone-shaped, truncate at the apex, bearing 5 oval stigmatic patches round the sides, obscurely 5-lobed at the apex. Fruit a globose, ovoid or narrowly ellipsoidal berry ; seeds immersed in green pulp.

Fruits more or less globose :
  Fruits up to 1·5 cm. in diameter ; flowers small,
    staminal tube 3·7 mm. long and 1·2 mm. in
    diameter ; leaves oblong-elliptic .   .   . 1. *W. stuhlmannii*
  Fruits up to 5 cm. in diameter ; flowers larger,
    staminal tube 4–5 mm. long and 2–3 mm. in
    diameter ; leaves elliptic to oblong-lanceolate . 2. *W. ugandensis*
Fruits ellipsoidal, elongated .   .   .   .   . 3. *W. elongata*

1. **W. stuhlmannii** *Engl.*, P.O.A. C : 276 (1895) ; Warburg in E. P. Pf. 3 (6) : 318 (1895) ; V.E. 3 (2) : 545, fig. 246 (1921) ; T.T.C.L. 107 (1949) ; K.B. 1954 : 542 (1955). Type : Tanganyika, Mpizi near Pangani, *Stuhlmann* 156 (B, holo.†). Neotype : Tanganyika, Msumbugwe, SW. of Pangani, *Gilchrist* 1 ! (EA, neo., K, isoneo.)

Tree 12–24 m. tall, glabrous ; bole rather short, clear of branches for 3 m. ; bark yellow to blackish-grey, splitting into irregular plates. Leaves oblong-elliptic, 3–9·5 cm. long and 1·4–3·2 cm. wide, often a little falcate or unequal

* The bracts, sepals and petals have also been termed sepals, inner and outer petals, respectively.

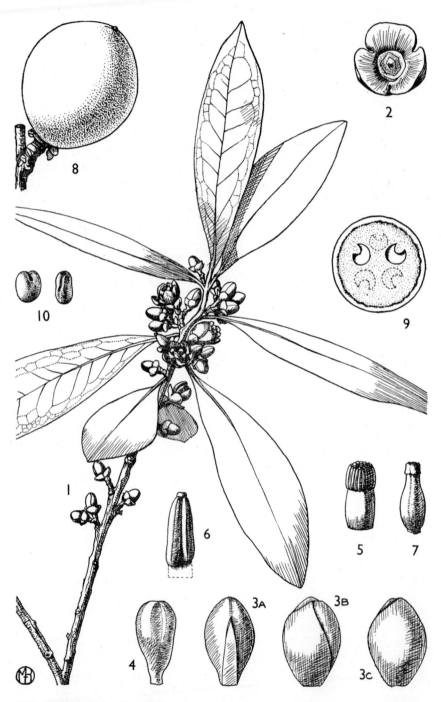

FIG. 1. *WARBURGIA UGANDENSIS* subsp. *UGANDENSIS*—**1**, flowering branch, × 1 ; **2**, bracts, × 4 ; **3**, sepals, × 4 ; **4**, petal, × 4 ; **5**, androecium, × 4 ; **6**, anther, × 10 ; **7**, pistil, × 4 ; **8**, fruit, × 1 ; **9**, cross-section of fruit (diagrammatic), × 1 ; **10**, seeds, × 1. 1–7, from *G. R. Williams* 495 ; 8–10, from *Verdcourt* 1010 & 1010A.

at the apex, very glossy and dark green above, paler beneath, acute at the apex, cuneate at the base and slightly involute ; venation darker green below, scarcely raised ; petiole 3–5 mm. long. Flowers solitary or in small cymes of 3–4, axillary. Bracts, 2·7 mm. long and 3 mm. wide, obscurely ciliate, green, ovate-reniform, very obtuse, only covering the flowers in very young buds. Sepals ovate, obtuse, green, 5·2 mm. long and 3·5–4 mm. wide ; petals spathulate, obtuse, yellow-green, 4 mm. long and 2 mm. wide. Staminal tube yellow-green, 3·7 mm. long and 1·2 mm. in diameter, prolonged 0·2 mm. beyond the anthers ; thecae 1 mm. long. The anthers are fully grown in bud and the tube elongates as the bud opens. Ovary 2·5 mm. long ; ovules 10. Style 0·5 mm. long ; stigmatic patches dark, becoming darker after fertilization. Fruit green with waxy bloom, ellipsoidal at first, globose when ripe, about 1·5 cm. in diameter. Seeds compressed, ± cordate, yellow-brown, 3–5 mm. long.

KENYA. Kwale District : Vanga, Ngoa, Nov. 1929, *Graham* 2208 !
TANGANYIKA. Pangani District : Msumbugwe, Nov. 1953, *Gilchrist* 1 ! & Feb. 1950, *Gane in A.H.* 9922 !
DISTR. **K**7 ; **T**3 ; apparently endemic in the coastal provinces of East Africa. It will almost certainly be found between the two localities mentioned
HAB. Coastal forest of the *Albizia-Ostryoderris* type and in open woodland, 200 m.

2. **W. ugandensis** *Sprague* in J.L.S. 37 : 498 (1906) ; I.T.U., ed. 2, 71, col. pl. 5 (1952). Type : Uganda, Toro, *Dawe* 510 ! (K, holo., ENT, iso.)

Tree up to 42 m. tall but often quite small (about 5 m.) in many Kenya localities, glabrous. Bark and habit similar to that of *W. stuhlmannii*. Leaves oblong-lanceolate, elliptic or oblong-elliptic, often a little falcate with costa eccentric, very glossy and dark green above, paler beneath ; venation sometimes a little more prominent beneath than in *W. stuhlmannii*, lamina acute at the apex, cuneate at the base and slightly involute, 3–15 cm. long and 1·4–5·0 cm. wide ; petioles 3–5 mm. long. Flowers solitary or in small cymes of 3–4, axillary. Bracts and flowers very similar in shape to those of *W. stuhlmannii*. Bracts thick, 3 mm. long and 3–3·5 mm. wide, ciliate. Sepals 6–7 mm. long and 4–4·5 mm. wide ; petals 5–7 mm. long and 2·5–3 mm. wide, obovate-spathulate, rarely with small lateral expansions at the middle. Staminal tube 4–5 mm. long and 2–3 mm. in diameter, thecae 1·5–2 mm. long. Ovary 2·6–4 mm. long ; ovules 25–30. Style 0·5–1 mm. long. Fruit at first greenish and ellipsoidal, later subspherical turning purplish, up to 5 cm. in diameter. Seeds compressed, ± cordate, yellow-brown, 1–1·5 cm. long.

KEY TO INTRASPECIFIC VARIANTS

Leaves elliptic to oblong-elliptic   .   .   .   .   subsp. **ugandensis**
Leaves oblong-lanceolate   .   .   .   .   .   subsp. **longifolia**

subsp. **ugandensis**

Tree to 42 m. tall. Leaves oblong-elliptic or elliptic, 3–15 cm. long and 1·4–5·0 cm. wide. Bracts 3 mm. long and 3·5 mm. wide. Sepals 6–7 mm. long and 4–4·5 mm. wide ; petals 5·5–7 mm. long and 3 mm. wide. Staminal tube 5 mm. long and 2–3 mm. in diameter, thecae 2 mm. long. Ovary 4 mm. long ; style 1 mm. long. Fig. 1.

UGANDA. Toro District : Kibale Forest, Aug. 1936, *Eggeling* 3131 ! & 3137 !
KENYA. Nairobi, July 1952, *G. R. Williams* 495 ! ; Kericho District : Sotik, June 1953, *Verdcourt* 960 !
TANGANYIKA. Bukoba District : Minziro Forest Reserve, Sept. 1950, *Watkins* 518 ! ; Lushoto District : Usambara Mts., Shume–Malindi road, Aug. 1950, *Greenway & Verdcourt* 8463 ! & Mar. 1946, *Silvana Abedi in F.H.* 1459 !
DISTR. **U**2 ; **K**4, 5, 6 ; **T**1, 3 ; Belgian Congo and the Transvaal
HAB. Lowland rain-forest, upland dry evergreen forest and its relicts in secondary bushland and grassland ; also on termite-hills in swamp forest, 1100–2200 m.

SYN.   *W. breyeri* Pott in Ann. Transvaal Mus. 6 : 60 (1918). Type : Transvaal, Pietersburg District, Drakensberg, near Macoutsie [Makoetsi] River, July 1917, *Breyer in Transvaal Mus.* 17573 ! (PRE, holo., K, iso.)

VARIATION. In the eastern part of its range (Bukoba–Toro) this subspecies is taller and has a rather different ecology. Field labels of specimens collected in this area always mention that the bark smells of sandalwood. The possibility of ecotypes should be borne in mind. It seems probable that the four taxa in this genus are the result of geographical and ecological isolation.

     subsp. **longifolia** *Verdcourt* in K.B. 1954 : 543 (1955). Type : Tanganyika, Lindi District, Rondo Plateau, *Bryce* 1 ! (EA, holo., K, TFD, iso.)

Tree to 27 m. tall. Leaves oblong-lanceolate, 7–12 cm. long and 1·7–2·1 cm. broad, costa rather eccentric. Bracts 3 mm. long and wide, ciliate. Sepals 6·5 mm. long and 4·5 mm. wide ; petals 5 mm. long and 2·5 mm. wide. Staminal tube 4–4·5 mm. long, thecae 1·5 mm. long. Ovary 2·6 mm. long, ovules 30. Style 0·5 mm. long.

TANGANYIKA. Lindi District : Rondo Plateau, Feb. 1951, *Eggeling* 6057 !
DISTR. **T**8 ; endemic in southern Tanganyika
HAB. Lowland rain-forest, 800 m.

NOTE. This subspecies is more distinct than leaf measurements alone indicate. The rather long parallel-sided leaves are quite different from leaves of similar length found in the other taxa. It is nearer in floral morphology to *W. ugandensis* than it is to *W. stuhlmannii* and there are not enough characters to distinguish it specifically. Further fruiting material is required.

3. **W. elongata** *Verdcourt* in K.B. 1954 : 544 (1955). Type : Tanganyika, Uzaramo District, Vikindu, *Eggeling* 6813 ! (EA, holo., K, iso.)

Small tree or evergreen shrub with 1–6 stems up to 6 m. tall, with rounded spreading crown ; branchlets ultimately drooping. Bark thin, grey-brown, smooth, with scattered cinnamon-coloured lenticels. Leaves papery, elliptic or oblong-lanceolate, slightly falcate, densely pellucid-punctate, acute at the apex, cuneate and narrowly winged at the base, lamina 4–15 cm. long and 2–5·5 cm. broad, minutely rugulose, margins slightly reflexed, petioles 3–7 mm. long. Flowers axillary, solitary or in 2-flowered cymes ; pedicels 2 mm. long. Bracts 2–3 mm. long and 3·2–4·5 mm. wide, very obtuse. Sepals convex, elliptic or obovate-spathulate, apically rounded, 4–6·5 mm. long and 2–4·5 mm. wide. Petals 3·5–6 mm. long and 1·5–2·5 mm. wide. Staminal tube 5 mm. long, thecae 2–2·5 mm. long. Ovary 4·5 mm. long, ovules ± 26. Fruit clavate, elongate, narrowed to the base, 5·5–6 cm. long and 2 cm. in diameter at the widest part, acute at the apex. Mature seeds unknown.

TANGANYIKA. Uzaramo District : Vikindu Forest, Sept. 1953, *Paulo* 157 ! & May 1954 (fl.), *Eggeling* 6813 ! & May 1954 (fr.), *Omari* 15 !
DISTR. **T**6; apparently endemic ; further collecting will probably show it to be more widely distributed
HAB. In undergrowth of coastal riverine forest of *Khaya*, *Macrolobium*, *Parkia*, *Afzelia*, *Trachylobium*, *Barringtonia* and *Uapaca* ; also on the edge of *Pandanus-Raphia* swamp forest, 30 m.

INDEX TO CANELLACEAE